Secrets to Raising Successful Backyard Chickens 2024

Updated & Expanded, Your Complete Chickenkeeping Resource

Arthur N. Robinson

Copyright ©2024, Arthur N. Robinson
All rights reserved.
No portion of this book may be reproduced in any form without written permissionfrom the publisher or author, except as permitted by U.S. copyright law.

Table of Contents

Part 1: Getting Started with Backyard Chickens

Introduction

Chapter 1: Is Raising Chickens Right for You? (Considering lifestyle, legalities, and expectations)

Chapter 2: Planning Your Flock: Size, Breeds (Egg Layers, Meat Birds, Dual-Purpose), and Age (Chicks, Pullets)

Chapter 3: Coop Dreams: Design, Construction Considerations (Safety, Ventilation, Space), and Essential Equipment (Feeders, Waterers, Nesting Boxes)

Chapter 4: Creating a Chicken Paradise: Run Fun (Fencing, Security from Predators), and The Great Outdoors (Free-Range vs. Confined)

Part 2: Keeping Your Flock Happy and Healthy

Chapter 5: Flock Feeding 101: Understanding Chicken Nutrition (Commercial Feed, Supplements, Scratch, Kitchen Scraps), and Watering Essentials

Chapter 6: Coop Care and Maintenance: Cleaning Schedules, Bedding Options, and Keeping Predators Out

Chapter 7: Keeping Your Chickens Healthy: Common Ailments, Preventative Measures, and Basic First Aid

Chapter 8: The Wonderful World of Eggs: Egg Collection, Storage, Understanding Egg Quality and Color Variations

Part 3: Beyond the Basics

Chapter 9: Going Green with Your Flock: Composting Chicken Manure, Sustainable Practices, and Eco-Friendly Coop Design

Chapter 10: Raising Chicks: Brooder Basics, Chick Care, and Integrating Them into the Flock

Chapter 11: Winter Woes? Keeping Your Chickens Warm and Happy During Colder Months

Chapter 12: Backyard Chicken FAQs: Egg-cellent Answers to Common Questions

Part 4: Bonus!

Chapter 13: Beyond Eggs: Processing Meat Birds for Home Consumption (Butchery Basics, Legality Considerations)

Chapter 14: Crafting a Backyard Chicken Community: Local Clubs, Online Resources, and the Joys of Sharing Your Flock

Conclusion

Part 1
Getting Started with Backyard Chickens

Introduction

Imagine gathering fresh eggs each morning, a picturesque coop nestled in your backyard, and hens contentedly clucking as they roam your garden. These are the delights of maintaining your own backyard flock. But how do you get started? Whether you're an experienced chicken keeper or a curious beginner, "Secrets to Raising Successful Backyard Chickens 2024: Updated & Expanded" is your ultimate guide to making your dream flock a reality.

This all-encompassing guide has been updated to incorporate the latest trends and best practices in backyard chicken keeping. We cover every aspect of the journey, from selecting the ideal breeds for your needs to constructing a secure and comfortable coop. You'll discover crucial advice on feeding your chickens a nutritious diet, ensuring their health and happiness, and collecting those delectable, homegrown eggs.

This edition goes beyond the fundamentals, exploring:

Innovative Coop Designs: Discover functional and stylish coop options tailored to your space and budget.

Biosecurity Measures: Learn how to protect your flock from diseases and predators.

Sustainable chicken keeping: Uncover eco-friendly practices to minimize your environmental footprint.

Solutions to common issues: Find effective strategies for challenges like feather pecking and molting.

"Secrets to Raising Successful Backyard Chickens in 2024 is your reliable partner on this fulfilling journey. We'll help you create a flourishing coop, enjoy the companionship of your hens, and benefit from fresh, homegrown eggs—all while fostering a connection with nature and sustainable living.

Raising backyard chickens isn't just about gathering eggs (though, let's admit, fresh eggs are a big perk!). It's about adding a touch of rural charm to your backyard. Watching your hens strut, preen, and dust bathe is a surprisingly soothing way to relax. Their gentle clucking and foraging bring life and tranquility to your outdoor space.

Backyard chickens are also relatively low-maintenance pets. Unlike dogs or cats, they need minimal training and are fairly independent. They also offer natural pest control, happily eating insects and grubs that might otherwise harm your garden. Their droppings, when composted properly, become nutrient-rich fertilizer for your plants, creating a beneficial cycle for both your coop and garden.

This updated edition of "Secrets to Raising Successful Backyard Chickens" reflects the growing interest in urban homesteading and sustainable living. As awareness of food security and environmental impact rises, keeping chickens provides a sense of self-reliance and a connection to your food source. It's a rewarding way to live more sustainably while enjoying the companionship of your feathered friends.

Dive into "Secrets to Raising Successful Backyard Chickens 2024, and discover the joys and practicalities of keeping your own flock. We'll equip you with the knowledge and confidence to embark on this exciting adventure!

Chapter One
Is Raising Chickens Right for You? Exploring Lifestyle, Legalities, and Expectations

The allure of fresh, homegrown eggs and a picturesque coop filled with lively hens is certainly enticing. However, before jumping into the world of backyard chickens, it's essential to assess your lifestyle, local laws, and what you can realistically expect. This guide will help you determine if raising chickens is the right choice for you.

Assessing Your Lifestyle

Space Requirements: Chickens need a secure coop and a spacious area to roam and forage. Evaluate your available space to ensure it can support a comfortable living area and an exercise-friendly run.

Time Commitment: Daily responsibilities include feeding, cleaning the coop, and collecting eggs. Consider whether you can consistently dedicate time to these tasks and periodically check on your chickens' health and safety.

Noise and Odors: Chickens, while endearing, can be quite noisy. Roosters crow, and hens cluck and chatter. Additionally, managing manure to control odors is necessary. Think about how this might affect your neighbors and your proximity to other homes.

Understanding Legalities

Local Regulations: Many municipalities have specific rules regarding backyard chickens, such as limits on the number of hens, coop size requirements, and restrictions on roosters. Thoroughly research local regulations before getting chickens.

HOA Rules: If you live in a community governed by a Homeowners Association (HOA), check their rules regarding backyard chickens. Non-compliance can lead to fines or other penalties.

Setting Realistic Expectations

Egg Production: While fresh eggs are a major benefit, don't expect a continuous supply. Hens are most productive in their first couple of years, with egg production decreasing over time. They may also stop laying during molting periods.

Ongoing Maintenance: Maintaining a clean and healthy coop is vital for your hens' well-being. This involves regular cleaning of bedding, droppings, and feeders. Be prepared for the ongoing effort required to keep the coop in good condition.

Predator Protection: Chickens are vulnerable to various predators. Ensuring their safety requires building a secure coop with predator-proof fencing.

Beyond the Basics

Raising chickens demands commitment, but the rewards are numerous. From enjoying fresh eggs to natural pest control and the simple pleasure of caring for these intriguing birds, the benefits are substantial. This guide serves as a starting point.

By carefully considering your lifestyle, understanding local laws, and setting realistic expectations, you can make an informed decision about raising chickens. Should you choose to proceed, resources like "Secrets to Raising Successful Backyard Chickens 2024: will provide the knowledge and confidence needed to establish a thriving coop and experience the joys of backyard chicken keeping.

Chapter Two
Planning Your Flock: Size, Breeds, and Age – Crafting Your Backyard Chicken Dream Team

Now that you've decided to take on the rewarding challenge of raising backyard chickens, it's time to plan your flock. This involves figuring out the ideal size, selecting the right breeds, and deciding on the appropriate age for your chickens.

Determining Your Flock Size

Several factors influence the optimal size of your flock:

Available Space: Your coop and run should provide enough room for all your chickens. Overcrowding can cause stress, disease, and aggressive behavior. Generally, plan for 4 square feet per chicken inside the coop and 8-10 square feet per chicken in the run.

Egg Requirements: If your main goal is to have fresh eggs, consider how many eggs a hen typically lays per week (about 4-6). Estimate your family's weekly egg consumption to determine the number of hens you need.

Local Regulations: Many municipalities limit the number of hens you can keep. Always check local laws before getting chickens.

Choosing the Right Breeds

Once you know how many chickens you can keep, it's time to pick your breeds.

Here are the main categories

Egg Layers: These breeds are known for high egg production. Popular options include Rhode Island Reds, Leghorns, and Australorps. They are usually smaller birds with friendly dispositions and moderate space needs.

These breeds are bred for meat production, growing quickly to a good size for consumption. Common choices are Cornish Cross and Jersey Giants. Note that meat birds are not good egg layers and need different care than egg-laying breeds.

Dual-Purpose Birds: These breeds provide both eggs and meat. Plymouth Rocks, Wyandottes, and New Hampshires are popular dual-purpose breeds. They are generally larger but lay fewer eggs compared to dedicated egg layers.

Deciding Between Chicks and Pullets

Your choice between chicks (baby chickens) and pullets (young hens) depends on your preferences:

Chicks: Raising chicks lets you watch them grow from a young age. However, chicks need a brooder with specific temperature and lighting conditions for the first few weeks and won't lay eggs for several months.

Pullets: Pullets are young hens, around 16-20 weeks old, that are close to starting egg production. They eliminate the need for a brooder and provide eggs sooner but are more expensive than chicks.

Tailoring Your Flock to Your Needs

Consider your lifestyle, goals, and available space when selecting your flock. If you want fresh eggs and have limited space, egg layers might be best. If you want chickens for both meat and eggs, dual-purpose breeds are a good choice. Starting with pullets gives you quicker access to fresh eggs but comes at a higher initial cost.

Going Deeper

"Secrets to Raising Successful Backyard Chickens 2024, offers more detailed information on specific breeds, their temperaments, care needs, and egg production rates. This resource will help you choose the best feathered friends for your backyard coop.

Remember, planning your flock is an exciting part of your chicken-keeping journey. By considering your space, needs, and resources, you can build a thriving and happy flock that will provide you with years of enjoyment and fresh eggs!

Chapter Three
Coop Dreams: Designing Your Backyard Chicken Haven

Your backyard chickens need a cozy, secure, and practical home. The coop is the heart of your flock's existence, so it's essential to plan and design it with care. This guide will help you create a coop that's both chicken-friendly and easy to maintain.

Designing Your Ideal Coop

Prioritize Safety: A secure coop is crucial to protect your chickens from predators. Use robust materials and ensure there are no gaps or weak points where animals could get in. Opt for predator-proof wire mesh for the run and bury the bottom edge underground to deter digging.

Ensure Proper Ventilation: Chickens produce a lot of moisture, making proper ventilation vital to prevent respiratory problems and ammonia buildup. Include vents in the coop walls, positioned high to avoid drafts. Adjustable vents can help you control airflow based on seasonal needs.

Provide Adequate Space: To prevent overcrowding, ensure your coop offers enough space for comfortable movement, roosting, and dust bathing. Allocate at least 4 square feet per chicken inside the coop. The run should provide at least 8-10 square feet per chicken for exercise and foraging.

Essential Equipment

With your coop design ready, it's time to outfit it with the essentials for your chickens:

Feeders: Choose appropriately sized feeders that minimize spillage. Hanging feeders or those with lids help reduce waste and keep food clean. Provide separate feeders for chicks or pullets, as they have different dietary needs from adult hens.

Waterers: Chickens require constant access to fresh, clean water. Automatic waterers are convenient, but it's wise to have backups in case of malfunctions. Waterers should be easy to clean and refill, and placed at a comfortable height for your chickens.

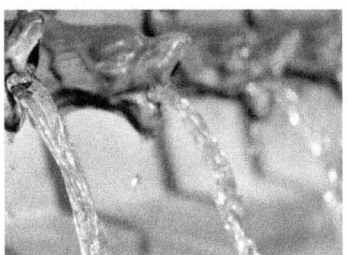

Nesting Boxes: Hens prefer laying eggs in a secluded, comfortable spot. Provide one nesting box for every 3-4 hens, filled with soft bedding like straw or wood shavings. Position them off the ground in a quiet corner of the coop.

Going Further

"Secrets to Raising Successful Backyard Chickens 2024: Updated & Expanded" offers detailed plans for coop designs, ventilation tips, and creative nesting box ideas. This resource will guide you through every step of the coop-building process, ensuring you create a functional and stylish home for your chickens.

Remember, a well-designed coop is an investment in your flock's health and happiness. By focusing on safety, ventilation, and space, you can create a coop that's both chicken-approved and easy to maintain, allowing you to enjoy the many rewards of backyard chicken keeping for years to come.

Chapter Four
Creating a Chicken Paradise: Run Fun and The Great Outdoors

Designing an ideal environment for your chickens means balancing security with freedom, ensuring your flock thrives while staying safe. Achieving this involves thoughtful planning of their run and understanding the pros and cons of free-range versus confined living.

Run Fun: Fencing and Predator Protection

The chicken run is a crucial part of their habitat, offering a space for exercise and exploration. Proper fencing is essential to create a safe and enjoyable run. A secure fence should be at least six feet high to prevent chickens from flying over and to keep predators at bay. Hardware cloth is preferred over chicken wire due to its durability and superior protection against predators like raccoons, foxes, and hawks. Burying the fence at least a foot underground helps to prevent predators from digging underneath.

In addition to secure fencing, a robust, predator-proof coop is vital. Elevate the coop off the ground to deter rodents and snakes, and ensure all doors and windows have secure locks. Motion-activated lights and alarms can provide extra security. Having a safe retreat for your chickens at night significantly reduces the risk of nighttime predation.

Enhance the run by adding features such as perches, dust baths, and varied terrain. Chickens love to scratch and peck at different surfaces, so incorporating sand, dirt, and mulch can make their environment more stimulating. Shade structures are also important to protect them from harsh weather.

The Great Outdoors: Free-Range vs. Confined

Choosing between free-ranging your chickens or keeping them confined depends on factors like space availability, local predator threats, and your personal preference.

Free-range chickens have the liberty to explore a larger area, forage for food, and exhibit natural behaviors, which can result in healthier, happier birds. They benefit from a varied diet, which can improve egg quality. However, free-ranging also increases their exposure to predators and the risk of wandering into dangerous areas. It requires more vigilance to ensure their safety.

Confined chickens, on the other hand, are easier to manage and protect. They are less likely to fall prey to predators or get lost. A well-designed run can still provide enough space and stimulation to keep confined chickens content.

However, they may miss out on the benefits of a more varied diet and natural foraging behaviors.

Ultimately, creating a chicken paradise involves understanding and addressing the needs of your flock. By providing a secure, enriching run and carefully considering the benefits and drawbacks of free-range versus confinement, you can create a safe and enjoyable environment where your chickens can thrive. Whether they roam freely or stay within a well-protected run, the key is to ensure they are healthy, safe, and happy.

Part 2
Keeping Your Flock Happy and Healthy

Chapter Five
Flock Feeding 101
Mastering Chicken Nutrition and Watering Essentials

Maintaining your flock's health and productivity hinges on providing a balanced diet and a steady supply of clean water. Understanding the essentials of chicken nutrition, including commercial feed, supplements, scratch, and kitchen scraps, is crucial for keeping your birds healthy and ensuring high-quality egg production.

The Basics of Chicken Nutrition

Commercial Feed

Commercial feed is the cornerstone of a chicken's diet, formulated to meet their nutritional needs at various life stages. Starter feed is high in protein to support the growth of chicks, while layer feed contains the calcium needed for egg production in mature hens. Choosing high-quality, well-balanced feeds ensures your chickens receive essential nutrients such as proteins, vitamins, minerals, and amino acids. For those interested in more natural feeding options, organic and non-GMO feeds are available.

Supplements

Supplements can enhance your chickens' diet, especially if they do not free-range. Grit is vital for digestion, as chickens use it to grind food in their gizzards. Oyster shell supplements provide extra calcium for strong eggshells. Probiotics and vitamins can boost the immune system and overall health, particularly during molting or periods of stress.

Scratch

Scratch is a mix of cracked grains like corn, barley, and wheat, offered as a treat rather than a staple. It encourages natural foraging behavior, providing mental stimulation and exercise. However, scratch is not nutritionally complete and should only constitute about 10% of their diet. Overfeeding scratch can lead to obesity and nutritional imbalances, so it's best given in moderation.

Kitchen Scraps

Kitchen scraps can diversify your chickens' diet and reduce waste. Suitable scraps include fruits, vegetables, grains, and cooked legumes. Avoid giving them anything spoiled, moldy, or toxic, such as avocado, chocolate, or raw beans. Introduce new scraps gradually to monitor their effects on your flock's health and prevent digestive issues.

Watering Essentials

Providing fresh, clean water is as crucial as proper feeding. Chickens can drink up to a pint of water daily, with intake increasing in hot weather. Dehydration can cause health problems and reduce egg production.

Regularly clean water containers to prevent algae and bacteria growth. Automatic waterers are convenient, ensuring a constant water supply. Place waterers at a height that prevents contamination from dirt and bedding. In winter, use heated waterers or frequently check to break the ice, ensuring water remains accessible.

During hot weather or stressful periods, adding electrolytes to the water can help maintain hydration and health. Apple cider vinegar is another beneficial addition, promoting gut health and preventing algae growth in waterers.

A comprehensive approach to feeding and watering your flock is essential for their health and productivity. By meeting their nutritional needs with a balanced mix of commercial feed, supplements, scratch, and kitchen scraps, and ensuring a constant supply of clean water, you create an environment where your chickens can thrive. Proper nutrition and hydration are the keys to a happy, healthy flock that will reward you with delicious eggs and lively companionship.

Chapter Six
Coop Care and Maintenance Cleaning Routines, Bedding Choices, and Predator Protection

Maintaining your chicken coop properly is vital for the well-being and safety of your flock. This includes regular cleaning, selecting suitable bedding, and implementing effective measures to deter predators. By staying on top of these tasks, you can create a safe and comfortable environment for your chickens.

Cleaning Routines

Keeping the coop clean is essential to prevent the accumulation of harmful bacteria, mites, and ammonia from chicken droppings. A good cleaning routine includes daily, weekly, and monthly tasks:

Daily Duties: Remove visible droppings from nesting boxes and around feeders and waterers. Ensure the water supply is clean and fresh, replenishing it as needed.

Weekly Duties: Perform a more thorough cleaning weekly. Replace bedding in high-traffic areas like nesting boxes and perches. Scrub feeders and waterers with warm, soapy water to prevent mold and bacteria buildup. Check the coop for any signs of damage or wear that could attract predators or cause weather-related issues.

Monthly Duties: Conduct a deep clean once a month. This involves removing all bedding and scrubbing the floors, walls, and perches with a disinfectant safe for chickens. Allow the coop to dry completely before adding fresh bedding. Regularly check for signs of mites or lice and treat the coop and chickens if necessary.

Bedding Choices

Selecting the right bedding is crucial for maintaining a clean and comfortable coop. Common bedding materials include straw, pine shavings, and sand:

Straw: Straw is popular due to its affordability and availability. It provides good insulation and is easy to replace. However, it can harbor mites and mold if not changed regularly.

Pine Shavings: Pine shavings are excellent for their absorbency and odor control. They help keep the coop dry and are easy to clean. Avoid cedar shavings, as their strong scent can harm chickens' respiratory systems.

Sand: Sand is gaining popularity due to its excellent drainage and ease of cleaning. It doesn't harbor pests and can be sifted to remove droppings, reducing the need for frequent changes.

Predator Protection

Protecting your flock from predators requires a combination of secure coop construction and vigilant maintenance.

Secure Construction: Ensure the coop is made of sturdy materials with no gaps larger than half an inch. Use hardware cloth instead of chicken wire, as it's more resistant to predators. Elevate the coop off the ground to prevent burrowing predators from accessing it.

Locks and Latches: Install robust locks and latches on all doors and windows. Raccoons, for example, are adept at opening simple latches. Padlocks or carabiner clips can provide an extra layer of security.

Perimeter Protection: Surround the coop with a predator-proof fence buried at least a foot underground to prevent digging. Consider adding an electric fence for additional protection.

Ongoing Maintenance and Monitoring: Regularly inspect the coop and fence for any signs of damage or weak points. Keep the area around the coop clean and free of debris that could provide cover for predators.

A well-maintained coop is crucial for a healthy and happy flock. By following a thorough cleaning routine, selecting appropriate bedding, and implementing robust predator protection measures, you can create a safe, comfortable, and hygienic environment for your chickens. These practices not only enhance the well-being of your birds but also ensure they remain productive and stress-free.

Chapter 7

Keeping Your Chickens Healthy Common Ailments, Preventative Measures, and Basic First Aid

Ensuring your chickens remain healthy is crucial for a thriving and productive flock. Familiarizing yourself with common illnesses, adopting preventive practices, and knowing basic first aid can greatly improve your ability to care for your birds.

Common Health Issues

Chickens can suffer from a range of health problems, including respiratory infections, parasites, and egg-laying complications.

Respiratory Infections: Diseases like Infectious Bronchitis, Mycoplasma, and Newcastle Disease can cause symptoms such as sneezing, coughing, and nasal discharge. These infections can spread rapidly among the flock.

Parasites: External parasites such as mites and lice, as well as internal parasites like worms, can lead to weakened chickens, decreased egg production, and poor overall health. Symptoms include feather loss, pale combs, and lethargy.

Egg-Laying Complications: Issues like egg binding, where an egg becomes stuck inside the hen, and prolapsed vent, where internal tissues protrude outside the body, can be severe and require immediate attention.

Preventive Measures

Preventive care is essential for maintaining a healthy flock.

Clean Environment: Regularly clean the coop and replace bedding to reduce the risk of bacterial and parasitic infections. Ensure proper ventilation to minimize ammonia buildup from droppings.

Balanced Nutrition: Provide a balanced diet with appropriate commercial feed, fresh greens, and occasional supplements. Always ensure access to clean, fresh water.

Routine Health Checks: Conduct regular health checks to detect issues early. Monitor for signs of illness, such as changes in behavior, appetite, or appearance. Check for external parasites and inspect droppings for internal parasites.

Vaccinations: Vaccinate your flock against common diseases. Consult a veterinarian to determine the necessary vaccinations for your area.

Biosecurity Measures: Limit contact with other poultry and wildlife to prevent disease introduction. Quarantine new birds for at least two weeks before integrating them into your flock.

Basic First Aid

Basic first aid knowledge can be crucial during emergencies.

Treating Minor Injuries: Clean cuts and abrasions with a saline solution and apply antiseptic. Keep the injured bird separate from the flock until healed to prevent pecking.

Managing Egg Binding: For an egg-bound hen, place her in a warm, humid environment to help relax her muscles. Gently massage the abdomen. Seek veterinary assistance if the egg does not pass within a few hours.

Handling Prolapse: For a prolapsed vent, clean the protruding tissue with warm water and apply a mild antiseptic. Gently push the tissue back into place and isolate the hen to prevent pecking. Consult a veterinarian for severe cases.

Addressing Respiratory Issues:

Isolate chickens showing respiratory symptoms to prevent the spread of infection. Provide a warm, stress-free environment and consult a veterinarian for appropriate antibiotics or treatments.

Maintaining your chickens' health requires a proactive approach. By understanding common health issues, implementing preventive measures, and being prepared with basic first aid skills, you can keep your flock healthy and productive. Regular monitoring and prompt action when problems arise are key to fostering a thriving chicken community.

Chapter 8
The Wonderful World of Eggs: Egg Collection, Storage, Understanding Egg Quality and Color Variations

Eggs are a beloved staple in diets around the world, prized for their versatility and nutritional value. For those who raise chickens, gaining insights into egg collection, storage, quality assessment, and color differences can significantly enhance the experience and efficiency of maintaining a flock.

Collecting Eggs

To ensure the best quality and productivity, regular egg collection is vital. Gathering eggs at least once a day, ideally in the morning, helps keep them clean and reduces the chances of damage. This practice also prevents hens from becoming broody and deters predators or spoilage. Use clean, dry hands or gloves to minimize contamination during collection.

Regularly check and clean nesting boxes, removing any debris or soiled bedding.

Storing Eggs

Proper storage is key to preserving egg freshness. After collection, store eggs in a cool, dry place. Refrigeration is optimal, as it can keep eggs fresh for up to three weeks. In the absence of refrigeration, a consistently cool environment like a cellar can suffice. Keeping eggs in their carton or a specific egg container protects them from absorbing strong odors and flavors from other foods. Store eggs with the pointed end down to maintain the integrity of the air cell, which helps preserve freshness.

Assessing Egg Quality

Several factors determine egg quality, including shell condition, yolk color, and the consistency of the egg white.

Shell Condition: A clean, unbroken shell indicates a healthy hen and proper handling. Thin or cracked shells may suggest dietary issues, such as insufficient calcium.

Yolk Color: The yolk's color is influenced by the hen's diet. A rich, deep orange yolk often results from a diet high in greens and insects, which provide xanthophylls and carotenoids. Pale yolks are more common in hens fed primarily on commercial feed.

Egg White Consistency: Fresh eggs have thick, gelatinous whites that hold their shape when cracked. Over time, the whites become more fluid and spread out, indicating decreased freshness.

Egg Color Variations

Egg color adds a delightful visual variety to egg collection. Chicken eggs come in several colors, from white and brown to blue, green, and speckled. The color of an egg's shell is determined by the breed of the hen and does not affect the egg's taste or nutritional content.

White Eggs: Laid by breeds such as Leghorns, white eggs are prevalent in commercial egg production due to the prolific laying capabilities of these hens.

Brown Eggs: Common among backyard flocks, brown eggs are laid by breeds like Rhode Island Reds and Orpingtons. The brown shell color results from pigments deposited as the egg travels through the hen's oviduct.

Blue and Green Eggs: Produced by breeds like Araucanas and Ameraucanas, the blue hue comes from the pigment oocyanin, which permeates the shell.

Understanding the nuances of egg management can enrich your experience as a chicken keeper. By mastering egg collection and storage techniques, recognizing high-quality eggs, and appreciating the diverse colors, you can fully enjoy the rewards of your flock's efforts. Whether you're a veteran poultry enthusiast or just starting, these insights will help you make the most of your egg production.

Part 3
Beyond the Basics

Chapter 9
Going Green with Your Flock: Composting Chicken Manure, Sustainable Practices, and Eco-Friendly Coop Design

Raising chickens presents an excellent chance to implement sustainable and eco-friendly practices. By composting chicken manure, adopting green methods, and designing an eco-friendly coop, you can lessen your environmental impact while boosting your flock's health and productivity.

Composting Chicken Manure

Chicken manure is a gardener's treasure, packed with nitrogen, phosphorus, and potassium. However, it needs proper composting to avoid plant damage and eliminate pathogens.

Collection: Collect manure from the coop and run, mixing it with bedding materials like straw or wood shavings. This combination of "greens" (manure) and "browns" (bedding) creates an ideal compost mix.

Composting Process: Build a compost pile in a well-ventilated area, ensuring it's at least 3 feet high and wide to maintain heat. Regularly turn the pile to aerate it and keep it moist. The compost should reach at least 140°F to kill harmful bacteria. After 6-12 months, the compost will become dark and crumbly, ready to enrich garden soil.

Application: Apply the finished compost to garden beds, providing nutrient-rich soil that supports healthy plant growth and reduces the need for chemical fertilizers.

Sustainable Practices

Incorporating sustainable practices into your chicken-keeping routine can significantly reduce waste and promote a healthier environment.

Feeding: Use organic or non-GMO feed to ensure natural ingredients for your chickens. Supplement their diet with kitchen scraps and garden waste, cutting down on food waste and offering varied nutrition.

Water Conservation: Collect rainwater in barrels to water your chickens. Ensure waterers are designed to minimize spillage and evaporation.

Integrated Pest Management: Opt for natural pest control methods instead of chemicals. Introduce beneficial insects, like ladybugs, and plant pest-repelling herbs around the coop.

Eco-Friendly Coop Design

An eco-friendly coop benefits both the environment and your chickens' living conditions.

Materials: Build the coop with sustainable or recycled materials. Reclaimed wood, recycled metal, and natural insulation like straw bales are great options.

Energy Efficiency: Maximize natural light and ventilation to reduce reliance on artificial lighting and climate control. Position the coop to take advantage of winter sunlight and summer shade. Consider adding solar panels to power necessary electrical components.

Waste Management: Use the deep litter method in the coop, allowing bedding to accumulate and decompose over time. This reduces waste and provides extra warmth during colder months. Periodically, add the litter to your compost pile.

Integrating green practices into your chicken-keeping routine involves composting manure, adopting sustainable feeding and watering methods, and designing an eco-friendly coop. These steps create a harmonious environment that benefits your chickens and the planet. Embracing sustainability not only reduces waste and conserves resources but also fosters a healthier, more productive flock, demonstrating that green living and successful poultry keeping go hand in hand.

Chapter 10
Raising Chicks: Brooder Basics, Chick Care, and Integrating Them into the Flock

Raising chicks can be a fulfilling endeavor, requiring meticulous care to ensure they mature into robust, productive hens. Mastering brooder setup, chick care, and integration into an existing flock is essential for successful poultry rearing.

Essentials of a Brooder

A brooder is vital for providing a warm and secure environment for chicks during their initial weeks.

Setting Up the Brooder: Brooders can range from large cardboard boxes and plastic bins to specially crafted brooder pens. They should have tall sides to keep chicks contained and ample ventilation to prevent overheating.

Heat Source: To replicate the warmth of a mother hen, chicks need a steady heat source like a heat lamp or brooder plate. The temperature should start at approximately 95°F during the first week and decrease by 5°F each subsequent week until the chicks are fully feathered at 6-8 weeks.

Bedding: Utilize absorbent materials such as pine shavings or straw to maintain a dry and clean environment. Avoid slippery surfaces like newspaper, as they can cause leg issues. Replace the bedding regularly to ensure hygiene.

Food and Water: Supply chick starter feed, which is rich in protein to support their rapid growth. Use shallow waterers to prevent drowning, and always keep the water clean and fresh.

Chick Care Guidelines

Proper care is crucial for the healthy development of chicks.

Feeding: Provide a high-quality chick starter feed until the chicks reach about 8 weeks of age. Chicks need constant access to food and water due to their high metabolic rates.

Health Checks: Regularly inspect chicks for signs of illness, such as lethargy, pasty butt (fecal matter sticking to the vent), or respiratory issues. Address any health concerns immediately by consulting a vet or experienced poultry keeper.

Socialization: Handle and interact with your chicks regularly to acclimate them to human contact. This makes them more docile and easier to manage as they grow.

Integrating Chicks into the Flock

Careful planning is necessary when introducing chicks to an existing flock to ensure a smooth transition.

Initial Separation: Keep chicks in a separate area where they can see and hear the older birds without direct contact. This helps both groups get accustomed to each other gradually.

Supervised Introductions: After a few weeks, start supervised interactions in a neutral area. Closely monitor these sessions to prevent bullying and ensure the chicks' safety.

Size Considerations: Wait until the chicks are similar in size to the adult birds before fully integrating them. This minimizes the risk of injury from pecking and helps establish a balanced pecking order.

Provide Hiding Spots: Ensure there are plenty of hiding places and escape routes for the younger birds within the coop and run to avoid aggressive behavior from older chickens.

Raising chicks successfully involves creating a secure, warm brooder environment, attentive care, and gradual integration into the flock. By following these steps, you can ensure your chicks grow into healthy, productive members of your poultry community, resulting in a cohesive and thriving flock.

Chapter 11
Winter Woes? Keeping Your Chickens Warm and Happy During Colder Months

As winter draws near, it's essential to prioritize the comfort and well-being of your chickens. Although chickens are generally resilient, they need extra care during colder months to stay healthy and productive. Here are some key strategies for keeping your flock cozy throughout the winter season.

Preparing the Coop

Properly preparing the coop is crucial to shield your chickens from the cold.

Insulation: Insulation is key to retaining heat in the coop. Consider using straw bales around the exterior and foam insulation boards inside. Ensure the coop is draft-free but maintains adequate ventilation to prevent moisture buildup, which can cause respiratory problems.

Bedding: Employ the deep litter method to maintain warmth. Start with a thick layer of pine shavings or straw and add fresh layers regularly. The decomposing bedding generates heat, creating a natural warming effect.

Windows and Doors: Seal windows and doors properly to keep cold air out. Open windows briefly during the day for ventilation, but close them before nightfall to retain heat.

Heating Options

While chickens can handle the cold, extremely low temperatures might necessitate additional heating.

Heat Lamps: If you use heat lamps, make sure they are securely fastened to minimize fire risk. Red heat lamps are preferred as they are less disruptive to chickens' sleep cycles.

Heated Perches: Heated perches can provide direct warmth during roosting. These are particularly useful in smaller coops where space heaters aren't feasible.

Heaters and Brooders: In extremely cold climates, consider safe, low-wattage coop heaters or brooders. Always adhere to manufacturer guidelines to prevent accidents.

Nutrition and Hydration

Maintaining proper nutrition and hydration is essential in winter to help chickens generate body heat.

High-Energy Feed: Increase the protein and fat content in their diet. Offering cracked corn or suet can provide extra energy. Ensure they have access to layer feed and additional grains to keep them energized.

Water: Prevent water from freezing by using heated waterers or changing the water regularly. Adequate hydration is crucial for digestion and maintaining body temperature.

Outdoor Activity

Chickens still need to go outside for exercise and mental stimulation, even during winter.

Sheltered Runs: Cover part of the run to protect it from snow and wind. Use tarps, clear plastic panels, or old windows to create windbreaks and allow sunlight to warm the area.

Pathways: Clear snow from the run and create pathways with straw or wood chips. This encourages chickens to venture outside and keeps their feet dry and warm.

Regular Monitoring and Care

Frequently check on your chickens to ensure they are coping well with the cold.

Health Checks: Look out for signs of frostbite, especially on combs and wattles. Applying petroleum jelly can help protect against frostbite.

Behavioral Observations: Watch for signs of stress or illness, such as lethargy, reduced egg production, or changes in appetite. Address any issues promptly to prevent them from worsening.

Ensuring your chickens stay warm and happy during winter involves a combination of proper coop preparation, heating solutions, enhanced nutrition, and regular monitoring. By taking these steps, you can keep your flock healthy and comfortable throughout the colder months, ensuring they are ready to thrive when spring arrives. These efforts not only protect their well-being but also support continued egg production and a harmonious flock.

Chapter 12
Backyard Chicken FAQs: Egg-cellent Answers to Common Questions

Raising backyard chickens is a gratifying hobby, but it can come with many questions, especially for those new to it. Here are answers to some of the most frequently asked questions to help you start off right and keep your flock healthy and happy.

1. How many chickens should I begin with?

Starting with 3 to 6 chickens is ideal for beginners. This number is easy to manage and provides enough eggs for a small family. Chickens are social creatures, so having a few helps them establish a natural pecking order, contributing to their overall happiness.

2. Which breed should I select?

Your choice of breed should align with your goals. For egg production, consider breeds like Leghorns, Rhode Island Reds, and Australorps. For dual-purpose breeds that are good for both eggs and meat, Orpingtons or Plymouth Rocks are excellent. If you want friendly, pet-like chickens, look into Silkies or Buff Orpingtons.

3. How much space do chickens need?

Each chicken requires at least 2-3 square feet of coop space and 8-10 square feet of run space. Adequate space prevents overcrowding, which can lead to stress, feather pecking, and disease. It's important to provide enough room for roosting, nesting, and foraging.

4. What should chickens eat?

Chickens need a balanced diet, including commercial feed that meets their nutritional needs. Layer feed is best for egg-laying hens, while grower feed suits young birds. They also enjoy kitchen scraps, greens, and grains. Always provide grit to help with digestion and oyster shells for calcium, especially for laying hens.

5. How often do chickens lay eggs?

Most hens start laying eggs at about 5-6 months of age. Typically, a healthy hen lays one egg every 24-26 hours, although this can vary by breed and age. Egg production might decrease in winter due to less daylight and during molting, when hens replace their old feathers.

6. Is a rooster necessary for hens to lay eggs?

No, hens do not need a rooster to lay eggs. Roosters are only needed if you want fertilized eggs for hatching. Without a rooster, hens will still lay unfertilized eggs, which are ideal for eating.

7. How can I protect my chickens from predators?

Securing your coop and run is essential. Use hardware cloth instead of chicken wire, as it is more durable and secure. Make sure the coop is well-constructed with no gaps or weak points. Lock doors and windows at night, and bury fencing at least 12 inches underground to deter digging predators.

8. How do I care for chickens in winter?

Chickens can handle cold weather if their coop is dry and draft-free. Use deep litter bedding for extra warmth, and provide plenty of fresh water, making sure it doesn't freeze. Increase their feed slightly to help them generate body heat, and check for frostbite on combs and wattles.

9. Can chickens be kept in urban areas?

Many urban areas permit backyard chickens, but regulations vary. Check your local ordinances for specific rules regarding the number of chickens allowed, coop requirements, and noise restrictions. Urban chicken keeping offers fresh eggs and a way to manage kitchen scraps naturally.

10. What should I do if my chicken gets sick?

Isolate the sick chicken to prevent disease spread and consult a veterinarian with poultry experience. Common signs of illness include lethargy, loss of appetite, abnormal droppings, and respiratory issues. Prompt care and attention can help your chicken recover.

Raising backyard chickens involves a learning curve, but with these answers to common questions, you'll be well-equipped to maintain a happy, healthy flock and enjoy a steady supply of fresh eggs. Whether you're just starting out or are an experienced keeper, understanding these fundamentals can help ensure a successful and enjoyable chicken-keeping experience.

**Part 4:
Bonus!**

Chapter 13
Beyond Eggs: Processing Meat Birds for Home Consumption (Butchery Basics, Legality Considerations)

Raising chickens for meat offers a practical and rewarding way to provide your family with fresh, homegrown poultry. It's crucial to understand the basics of butchery and the relevant legal considerations to ensure a humane, efficient process while complying with regulations.

Butchery Basics

Processing meat birds at home requires careful preparation and a respectful approach to the animals.

Preparation

Fasting: Withhold food from the chickens for 12-24 hours before processing to clear their digestive tracts. Ensure they have access to water to stay hydrated.

Equipment: Gather necessary tools such as a sharp knife, a killing cone or restraining device, a scalding pot, plucking tools (either a mechanical plucker or hand plucking tools), and a clean workspace with tables and buckets for waste disposal.

Processing Steps

1. **Humane Slaughter:** Use a killing cone to restrain the bird, minimizing stress and injury. Make a swift, clean cut to the carotid arteries and jugular veins for a quick, humane death, allowing the bird to bleed out completely.

2. **Scalding and Plucking:** Submerge the bird in water heated to 145-150°F for 30-60 seconds to loosen feathers, making plucking easier. Remove feathers with a mechanical plucker or by hand.

3. **Evisceration:** Make a small cut at the vent and carefully remove the internal organs, ensuring not to puncture the intestines. Save edible organs like the heart, liver, and gizzard, and dispose of or compost the rest of the waste.

4. **Chilling:** Immediately chill the carcass in ice water to lower its temperature and prevent

bacterial growth. Keep the bird in the ice bath for at least 30 minutes, then transfer it to a refrigerator or freezer for storage.

Legal Considerations

Understanding and adhering to local, state, and federal regulations is essential when processing chickens for home consumption.

Local and State Regulations

Zoning Laws: Verify local zoning ordinances to ensure home processing of poultry is allowed in your area. Some residential zones may have restrictions on animal slaughter.

Health and Safety Codes: Adhere to state guidelines for sanitation and waste disposal. Use a clean, designated area for processing and manage waste materials properly to comply with health regulations.

Federal Regulations

Exemptions: The USDA provides exemptions for small-scale poultry producers processing birds for personal use or direct-to-consumer sales. For instance, the "1,000 bird exemption" allows producers to process up to 1,000 birds per year without USDA inspection, provided the meat is sold directly to consumers and not through commercial channels.

Labeling: If selling processed poultry, ensure proper labeling that includes the producer's name, address, and a statement indicating the exemption from USDA inspection.

Processing meat birds at home involves a blend of skill, respect for the animals, and adherence to legal requirements. By mastering the basics of butchery and understanding the regulations governing home processing, you can provide your family with high-quality, homegrown poultry while ensuring a humane and efficient process. This sustainable approach to meat production enhances self-sufficiency and deepens your connection to the food you consume.

Chapter 14
Crafting a Backyard Chicken Community: Local Clubs, Online Resources, and the Joys of Sharing Your Flock

Raising backyard chickens can be a fulfilling hobby, and sharing the experience with others makes it even more enjoyable. Building a community around your flock through local clubs, online resources, and social sharing can enhance your knowledge, provide support, and spread the joy of chicken keeping.

Local Clubs

Joining a local chicken club can connect you with other poultry enthusiasts.

Networking and Support: Local clubs provide a platform for exchanging knowledge, experiences, and advice. Whether you're new to chicken keeping or a seasoned pro, you can gain insights into best practices, solve common problems, and explore new ideas.

Events and Meetups: Clubs often host events such as farm tours, chicken swaps, and workshops. These gatherings offer opportunities to see different setups, learn from experts, and sometimes trade or purchase birds and supplies.

Community Projects: Participating in initiatives like school programs, community gardens, and local fairs can promote sustainable living and the benefits of backyard chickens.

These projects help foster a sense of community and shared purpose.

Online Resources

The internet offers a wealth of information and support for backyard chicken keepers.

Forums and Social Media Groups: Sites like Backyard Chickens, Reddit, and Facebook host numerous groups and forums where you can ask questions, share stories, and find advice on everything from coop design to health issues. These communities are active and responsive, providing real-time assistance.

Educational Websites and Blogs: Websites like The Chicken Chick, Fresh Eggs Daily, and the USDA's site offer comprehensive guides, articles, and how-to videos covering all aspects of chicken keeping. These resources are invaluable for both beginners and experienced keepers.

YouTube Channels: Channels dedicated to poultry farming, such as Justin Rhodes, Living Traditions Homestead, and Gold Shaw Farm, provide visual and practical guidance. Watching experienced keepers can help you learn new techniques and get inspired by their setups.

The Joys of Sharing Your Flock

Sharing your chicken-keeping journey with others brings unique joy and fulfillment.

Education and Inspiration: Inviting friends, family, and neighbors to visit your flock can inspire them to start their own backyard chicken adventure. You can share your knowledge about sustainable living, animal care, and the benefits of fresh eggs.

Community Bonding: Hosting informal gatherings or potluck dinners featuring dishes made with your fresh eggs creates a sense of community. Sharing your flock's produce with neighbors not only spreads goodwill but also showcases the rewards of chicken keeping.

Social Media Sharing: Documenting your chicken-keeping journey on social media platforms like Instagram, YouTube, or a personal blog allows you to connect with a broader audience. Sharing photos, videos, and stories can build a following of like-minded individuals and provide a platform for exchanging ideas and support.

Building a community around backyard chickens through local clubs, online resources, and social sharing enriches the experience of raising chickens. These connections provide invaluable support, foster community spirit, and promote the joys of sustainable living. By engaging with others who share your passion, you enhance your knowledge and contribute to a growing movement of backyard poultry enthusiasts.

Conclusion

Keeping backyard chickens is more than just a pastime; it's a multifaceted endeavor that enhances lives, promotes sustainability, and fosters community ties. From setting up a brooder and caring for chicks to processing meat birds and engaging with local chicken communities, each facet of poultry keeping presents distinct rewards and challenges.

Starting with the basics, raising chicks involves meticulous attention to brooder setup, temperature regulation, and proper nutrition. As chicks mature, transitioning them to an outdoor coop and integrating them into an existing flock must be done carefully to ensure a peaceful environment. Regular coop maintenance, including consistent cleaning and appropriate bedding choices, helps maintain a healthy living space and prevent common health issues.

Winter presents unique challenges, but with adequate coop insulation, heating solutions, and enhanced nutrition, chickens can thrive even in cold weather. Adopting sustainable practices, such as composting chicken manure and building eco-friendly coops, benefits your flock and positively impacts the environment. These efforts emphasize the importance of minimizing waste and promoting a self-sufficient lifestyle.

For those raising chickens for meat, understanding butchery basics and adhering to local and federal regulations ensures a humane and legal process. Proper equipment, humane slaughter techniques, and knowledge of USDA exemption guidelines allow backyard keepers to process their poultry efficiently and ethically. This aspect of chicken keeping highlights the complete cycle, from egg production to meat processing.

Creating a backyard chicken community significantly enhances the experience. Local clubs offer networking opportunities, support, and educational resources, while online forums, blogs, and YouTube channels provide a wealth of information and virtual camaraderie. Sharing your journey through social media or local events spreads the joy of chicken keeping and inspires others to start their own poultry adventures.

Egg production is a core benefit of keeping chickens. Understanding egg collection, storage, and recognizing quality and color variations adds to the practical joys of this endeavor. Fresh eggs from your flock are not only a delicious reward but also a testament to your care and commitment.

In summary, backyard chicken keeping is a richly rewarding practice that integrates animal care, sustainability, and community engagement. By mastering various aspects of poultry keeping—from raising chicks to processing meat birds, winter care, and community building—you create a holistic and fulfilling experience. Whether motivated by the desire for fresh eggs, sustainable living, or the simple joy of caring for animals, backyard chickens offer endless opportunities for learning and growth. This journey enhances your own life and contributes to a broader movement of conscious, connected, and environmentally responsible living.